AF615347

GRASSLANDS

BIOMES OF NORTH AMERICA

Lynn M. Stone

Rourke
Publishing LLC
Vero Beach, Florida 32964

www.rourkepublishing.com

PHOTO CREDITS: All photos © Lynn M. Stone

Title page: *Bison graze on prairie grasses in the shadow of badlands in Theodore R. Roosevelt National Park, North Dakota.*

Editor: Frank Sloan

Cover and interior design by Nicola Stratford

Library of Congress Cataloging-in-Publication Data

Stone, Lynn M.
Grasslands / Lynn M. Stone.
p. cm. — (Biomes of North America)
Summary: Looks at the grasslands of the United States, animals that dwell in them, and how grasslands are changing.
Includes bibliographical references (p.) and index.
ISBN 1-58952-685-6
1. Grassland ecology—Juvenile literature. 2. Grasslands—Juvenile literature. [1. Grasslands. 2. Grassland ecology. 3. Ecology.] I. Title. II. Series: Stone, Lynn M. Biomes of North America.
QH541.5.P7S744 2003
577.4—dc21

2003005102

Printed in the USA

CG/CG

Table of Contents

The Grasslands	5
The Grassland Community	11
Grassland Animals	14
Changing Grasslands	19
Glossary	23
Index	24
Further Reading/Websites to Visit	24

Canada
United States
Mexico
Cuba
North American Grassland
0
1500
KM
0
1000
Miles
Grassland

The Grasslands

When westbound pioneers left the Eastern forests, the open, nearly treeless grasslands greeted them. This was the grassland of the Midwest known as the tall-grass prairie.

Grasslands are covered with wild grasses, flowers, and other plants. In fact, the North American grasslands are habitats, or homes, for hundreds of different plants.

Fields and meadows are open grasslands, too. But most of their plants are not **native**. Instead they are **exotics**. Exotics come from other countries. Some are crop grasses, which people have planted.

Over hundreds of years, the prairie grasslands developed across much of North America's middle. Fire, wind, bison, and the dense roots of prairie plants helped keep trees from taking root.

The dense roots of grasses help keep the prairie free of invader plants, such as weeds and trees.

Three basic kinds of prairies developed. Short-grass prairies were furthest west, closest to the Rocky Mountains. Tall-grass prairies developed furthest east, where rainfall was greatest. Between the short and tall types lie mixed-grass prairies. Other types of grasslands developed further to the Northwest and in the dry Southwest.

Bison wander across a broad mixed-grass prairie in South Dakota.

The Grassland Community

The plants and animals of the grasslands live together in a natural community. Plants grow, in part, by making food from sunlight. Grassland plants become the food base for grassland animals.

Some grassland **predators**, like coyotes and badgers, eat meat. But they still depend upon plants because their **prey** is most often a plant-eater!

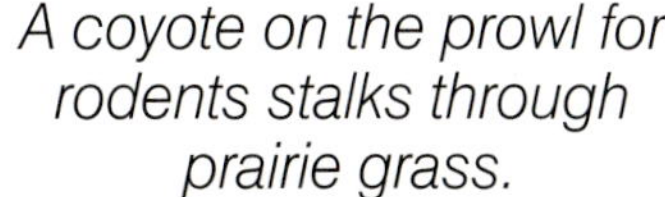

A coyote on the prowl for rodents stalks through prairie grass.

The mouse or ground squirrel a coyote eats, for example, lives on a diet of plants. A few grassland animals, such as the ornate box turtle, live on both plants and animals.

Grassland plants also provide shelter and hiding places for animals. And tall grasses protect smaller prairie plants from the harsh sun and wind.

The pronghorn antelope is one of the largest plant-eaters on the prairie.

Grassland Animals

Forest animals tend to be climbers. Many grassland animals are runners, like pronghorn antelope and mule deer, or diggers, like box turtles and badgers.

The best known prairie digger is the prairie dog. Prairie dogs are really a kind of ground squirrel. Tree squirrels climb. Ground squirrels dig burrows.

Prairie dogs nibble at grassland plants to fill their stomachs.

By nibbling away at the grassland plants, prairie dogs get both food and an open view to guard against predators.

American bison are perhaps the best known grassland animal. Sometimes called buffalo, American bison once numbered in the many millions. In the wild, they now live only in a few prairie **preserves**.

A bison bull stands in the grass of a North Dakota prairie.

Changing Grasslands

Grassland plants and animals are designed to survive the changes in climate. Most birds, for example, avoid prairie winters by **migrating** south each autumn. Other animals burrow underground. Some, like the bison, have heavy coats for winter. They shed their heavy coats each summer.

Extreme, but normal, differences in **climate** make natural changes in the grasslands each year. Winters are cold, harsh, and snowy. Summers are hot and dry.

Blazing-stars bloom on an Illinois tall-grass prairie preserve in early September.

The grasslands survived fire, wind, and extreme climate for thousands of years. But the rich eastern prairies were plowed to become the cropland of the Midwest. The western grasslands became ranchland for sheep and cattle. Few native prairies exist today.

A few, tall-grass prairies are being **restored**, however. People are planting prairie plant seeds and bringing some small prairies back to life.

Cattle ranching has become popular on the western grasslands.

A burrowing owl looks for a meal of insects.

Glossary

climate (KLY muht) — the weather of an area over a long period of time

exotics (egg ZAHT ickz) — plants or animals released into habitats of which they are not a natural part; especially plants or animals released into the wild from another country

migrating (MY grayt ing) — in the act of making a lengthy journey from one place to another, usually at the same time each year

native (NAYT iv) — a plant or animal that is a natural part of its habitat

predators (PRED uht uhrz) — animals that kill other animals for food

prey (PRAY) — an animal that is killed by another animal for food

preserves (pree ZERVZ) — places that are set aside for the protection of the plants and animals that live within it; refuges or sanctuaries

restored (ree STORD) — having been made back into what it used to be

INDEX

bison 16, 19
climate 19
coyote 13
exotic plants 6
grassland animals 14
grassland plants 11, 13
ground squirrels 14
habitats 5
migrating 19
native plants 6
Northwest 9
prairie dog 14, 16
prairies 5, 6, 9, 16
predators 11, 16
prey 11
Rocky Mountains 9
Southwest 9

Further Reading

Gray, Susan. *Grasslands*. Compass Point Books, 2001
Johnson, Rebecca L. *Walk in the Prairie*. Carolrhoda, 2001
Wilkins, Sally. *Grasslands*. Bridgestone Books, 2001

Websites To Visit

www.bellmuseum.org/mnideals/prairie/
www.yahooligans.com/Science_and_Nature

About The Author

Lynn Stone is a talented natural history photographer and writer. Lynn, a former teacher, travels worldwide to photograph wildlife in their natural habitat. He has more than 500 children's books to his credit.